A to Z
Magic Mushrooms making your own

for Total Beginners
lisa bond

Table of Contents

Copyright © 2017 by SB books..3
Introduction...4
Spores..7
    Spore Printing..8
    Strains..11
Preparing a Liquid Culture..14
    Jars...17
    Honey Water..18
    Pressure Cooker..20
    Injecting the Spores..22
    Here Comes the Mycelium...23
Growing With Brown Rice Flour Cakes..29
    Jar Lids..31
    Pressure Cooker..32
    Spawning Substrate..33
    Sterilization of the Jars 35
    Adding the Liquid Culture...36
    The Colony Grows..39
    Rest In Peace? (Contamination)...40
    Fruiting Chamber..41
    Consolidation...48
    Fruit(ing) Of Your Labor..50
    Mushrooms Are Blooming...54
    Fruiting Gone Bad..56
    Harvested… Now What?..57
    Consumption and Final Thoughts...60

A to Z
Magic Mushrooms making your own

for Total Beginners

lisa bond

Copyright © 2017 by SB books

All rights reserved. No part of this publication may be reproduced, distributed, or transmitted in any form or by any means, including photocopying, recording, or other electronic or mechanical methods, without the prior permission of the publisher, except in the case of brief quotations embodied in critical reviews and certain other noncommercial uses permitted by copyright law.

The information provided here is for educational purposes only, and the author and publisher take no liability for your actions or the results of using this information for your own research. It is strongly encouraged to review local, state, and federal law concerning the legality of any type of botanical or substances.

Introduction

In this offering, I will provide one method of growing mushrooms. It is a method that has been popularized by the likes of famous growers and forefathers like "Rogerrabbit" (Marc R. Keith) and tweaked throughout the years by the many others that have come after him to help put together useful resources for the growth and consumption of magic mushrooms. The processes outlined here are not vastly different than their methods (also referred to as TEK-style tutorials), but I have attempted to simplify the language for newcomers, explain some of the information that isn't absolutely critical for growing but helps to understand the growing process, and offer some slight updates when it seemed worthwhile. The end goal is a stepping stone, a starting point, a guide that is comprehensive enough to get you from start to finish yet short enough to keep your attention.

It is important, of course, to note that you need to take the time to do your own research as well. Not only is there a vast wealth of knowledge and experienced growers you can learn from, but there are also some legal considerations that may be of concern to you. The author (that's me) and the publisher take no responsible and absolve ourselves of all liability. Your actions are your own, and it is highly suggested that you take the time to understand local, state (or provincial), and federal laws that apply to you when growing, consuming, and potentially distributing magic mushrooms.

The book will be broken into three simple sections:

1. **Spores** – a brief introduction on mushroom spores, a small number of strains, spore printing, and spore syringes.

2. **Liquid Culture** – a chapter on how to begin the process of going from spores to a liquid culture, which is used for growing the actual mushrooms.

3. **Growing** – the brown rice flour cake method of growing mushrooms using a liquid culture.

4. It is highly advised to read all of the steps prior to starting. Not only will these help you understand the materials needed, but by understanding the steps in order before starting any of the steps, you will be better prepared and will better understand the reasoning behind these suggestions.

Lastly, I encourage you to have fun. Growing takes patience, but the entire process can be a fascinating adventure in science, spirituality, and intoxication. Your adventure may be riddled with some potential failures, and you can easily allow it to become a frustrating mess or a growing experience for yourself. It is truly up to you. If you continue to practice and attempt to learn from your mistakes, there is no reason you cannot enjoy the process even before you are able to consume your homegrown magic mushrooms. So please, try to make the most of it!

Spores

Before you can grow anything, you will need the spores required for the type of mushrooms you wish to grow. These spores are typically going to be stored in a syringe that mixes the spores with water. Using this syringe method, you will be first growing what is known as a **liquid culture**, which is the mycelium that is floating in water. **Mycelium** is the vegetable part of a fungus, consisting of branching white filaments known as the **hyphae**. Once your liquid culture is prepared, you will move on to growing mushrooms in what is known as a **fruiting chamber**. We will discuss this in great detail as the guide continues.

When you obtain spores for a spore syringe, they will either come as a **spore print,** the powdery substance that comes from a mushroom and falls to a surface below, or you will have a **prepared spore syringe** from a supplier. For the budding botanist, it is ideal to procure a spore syringe that is already prepared. However, **spore printing** is worth discussing, as you will likely want to go down this route later if for no other reason than curiosity.

Spore Printing

If you are not purchasing a prepared spore syringe for the process of creating a liquid culture, then you will need to work with spore printing in order to collect the spores and create your own spore syringe. This process can be done on homegrown mushrooms or wild mushrooms, but be sure that you're using a strain that you are familiar with, as growing the wrong mushroom and then ingesting can be a huge mistake. For most purposes, using homegrown mushrooms will produce the best results.

The process of spore printing is fairly simple, but you must maintain a sterile environment to achieve usable spores. With a glove box (a box that is devoid of air flow with holes for your arms to reach into) with lockable holes, you can place a sheet of paper or aluminum foil within the box, and taking only the caps of fresh mushrooms you want to gain spores from, you place them on the paper or aluminum foil

(foil may be easier to work with). Laying a sheet of paper over the cap is also useful as it can help avoid drying out the mushroom during the printing period. You must close the holes to your glove box for the best results. For most mushrooms, 16-24 hours is plenty of time to obtain a good amount of spores. When you remove the paper and the mushroom caps, a design resembling the mushroom cap's bottom should be apparent. This will be a collection of spores that can be used to create a spore syringe.

Again, sterilization is critical here. Introducing any contaminants can ruin your spores and make it impossible to create a liquid culture that will bear the results you're after. You'll need some sterilized distilled water and an **inoculating loop** or something with rounded edges to stir your spores into the water. Avoid anything with sharp or square edges for this, as this can damage your spores. Make sure to sterilize the inoculating loop or similar tool first, and then put a small amount of the sterilized distilled water on the spore print (hence the use of aluminum foil), stir with the inoculating loop, and then draw the spore-water concoction into a syringe. You can add more water afterwards. This keeps the spores in the water rather than on the edges of your plastic syringe.

That's really all there is to the process. I suggest that we skip this step for beginners for a couple of reasons. First, you most likely don't have access to a fresh mushroom that has been grown indoors until after you've grown your first flush. Once you have grown a high-potency mushroom that meets your desired results, you can use this method to hopefully continue growing the same mushrooms again with similar results. The other reason is that sanitation is huge deal when dealing with spore printing. Should any type of contaminant make its way into your spore printing or the needle, there's a good chance you'll be working with an entire batch of spores that are going to be less than desirable.

I encourage you to eventually try this method, but as a beginner, purchasing a spore syringe that is already prepared will make the process that much simpler.

Strains

There are many varieties of spores and fungus, but the most common starting point is going to be psilocybe cubensis. Within the psilocybe cubensis family are many **strains**. There is a popular saying within the shroom community that, "a cube is a cube," which suggests that they will all hold the psychotropic properties that magic mushrooms are known for. While psilocybe cubensis may not be the strongest variety of magic mushroom, it is relatively easy for the newcomer to grow, and most people are fine with their potency.

That said, a very small number of common strains available within the psilocybe cubensis family include:

1. **A** – The "classic" cubensis strain.

2. **Albino A+** — Not a truly albino fungus, but a mutation of the A strain. Not known for excellent growth, but excellent when

crossbreeding different strains.

3. **B+** — In the commercial world of magic mushrooms, B+ is often a go-to strain. It is known for creating large fruits, and while it grows slower than many commercialized strains, it is also known to be very forgiving, making it ideal for a beginner. This is our suggestion if you have any doubts on which strains you wish to start off with.

4. **Cambodian** — A highly revered strain for high yield and large fruits, it is said to be one of the more potent choices and has become a very popular choice amongst repeat cultivators.

5. **Golden Teacher** — The reviews on Golden Teacher are so mixed that one really must grow and experience this strain for themselves to decide if it interests them. The yield is not as fruitful as others, but some claim that the potency of these particular mushrooms is extremely impressive.

Of course, there is a huge variety of cubensis strains that we haven't remotely touched on, and it may be worth your while to investigate other strains. If this your first time, though, it is highly encouraged that you try to grow the B+ strain for its forgiving nature and large yields. Again, all cubes are going to have the desired effects when cultivated properly. The major differences tend to be the grow times and their appearances, though there is believed to be nuance between their psychotropic effects during experiments conducted by cultivators.

Again, it is highly advised to obtain a spore syringe that is already prepared for you. If you do want to work with spore printing and creating your own syringe instead, then there's no reason you shouldn't, but . From here on out, we are going to assume you're using a spore syringe that is already prepared, either through spore printing or by purchasing a prepared syringe.

Preparing a Liquid Culture

Since we have a prepared spore syringe, we can jump right into creating our liquid culture. There are several methods that can be used to achieve this liquid culture preparation. It is debatable, but the method using a honey-water liquid culture is often the easiest for newcomers. That said, the results are not as easy to control as more advanced methods. The major issue with the honey-water liquid culture method is that it is more likely to see contamination in the jars used to keep the mushrooms isolated from the outside world. As your skills improve, you can move on to other methods. For the beginner, the ease of setup is worth the risk.

With the honey-water method, you should expect at least half of your jars to become fruitful and remain without contamination. This method is typically faster and less involved as well, so it's perfect if you aren't ready for more serious effort to grow large amounts of mushrooms. To begin, you will need some supplies.

1. **First, you will need some large jars.** At a minimum, these should be half a liter, but larger jars are preferred. Make sure that these jars will seal properly. A good mason jar will do the trick, and it is suggested over jars that only have the single lid piece. Make sure the jars and lids are perfectly clean and sanitized.

2. **It is also helpful to have some syringes with 14- or 16-gauge needles.** Smaller needles are not ideal because the liquid culture of spores is going to be a thicker consistency than water. It doesn't pay to gum up your needles. When purchasing a spore syringe from a supplier, it will almost always be a 16-guage needle.

3. **You will require silicon** that is rated to withstand temperatures around 600 degrees Fahrenheit (roughly 300 degrees Celsius). An easy way to find this is to look for "gasket maker" silicone that is commonly used for automobiles, as these silicones are designed specifically for high amounts of height.

4. **A decent pressure cooker.** There are methods without pressure cookers, but typically for the best results, a pressure cooker is going to be far simpler.

5. **Liquid honey** for your honey-water liquid culture. Organic honey tends to work better than non-organic, and honey that is thin and runny is ideal in comparison to thicker honey. You will need about 4cc (4 milliliters), which you will be able to measure with measuring spoons, cups, or other methods.

6. **Marbles**, which will be added to your jar. You only need a few per jar, so there's no need to get a huge amount of these.

7. **Ethyl-alcohol lamp,** which is an easy method to help you sterilize your needles throughout the process. This can be purchased, or it can be made, but they are typically inexpensive so there's no need to go through the process of building your own unless you are just interested in it.

Jars

Now that you have gathered your supplies, the first step is to get your jars prepared for the rest of the process.

To start, puncture the jar lid with a hole that is sized for your syringe needle to push through easily. Any method of creating this hole is fine, but a large tack or nail is fine. When creating this hole, it is ideal that you push from the top of the lid downwards, so any sharp edges are within the jar and not sticking out of the top.

Take the silicone, and apply it to the hole and the area around the hole to create a seal. Within the hole itself, you may need quite a bit more silicone than the area around the hole you've created. When creating this seal, it is important that the silicone cannot break during the stage where we place the jars in a pressure cooker for sterilization. It's fine if this is not neat and clean, and thicker is better. Here comes the first of many waiting stages in your new botany project. You need to wait 24 hours for the silicone to harden. Note that harden silicone will still be somewhat pliable. You are not looking for rock hard consistency. Some cultivators will allow their pressure cooker to quicken the hardening process instead, but I find it better to let it harden the way it is was meant to be hardened. So let it sit! This whole process is a test of patience, but the results are going to be well worth it.

Honey Water

To prepare your honey water for the liquid culture, start by boiling clean water. If you haven't already, clean your marbles with alcohol, rinse them in sterilized water, and place them in the sterilized jars, as these are required for proper stirring later. You will need enough water to fill your jars and leave roughly 5cm of air on the top. When mixing your honey into your water, you should mix in an exact 4% ratio, so every 96ml (96cc) of water needs exactly 4ml (4cc) of honey. When mixing in the honey, make sure that it dissolves as much as possible.

It is paramount that the rim of the jar doesn't get touched by any water. The area around the lid and the jar should be completely dry, as water in this area is an easy way to bring in contamination from the outside world. It is ideal to take the time to clean off the rubber lining and rim of the jar with alcohol swabs and allow it to dry. This will help to remove any potential oils that can cause contamination or stop the jar from sealing properly. Next, you'll simply twist the lid tightly on the jar.

When the lid is tight, we need to create an access point for oxygen that doesn't allow for much else to come into the jar. The easiest way to do this is to give the jar lid a twist of about $1/4^{th}$ the way around. We need this oxygen allowance because the jar is going to be sterilized in the pressure cooker before we continue. If there is not an allowance for air, the pressure that will build up inside the jar may cause it to crack or otherwise become damaged. Then all this work is for nothing.

Alternatively, you can place a needle from a syringe through the silicone-covered hole with a bit of cotton over it. This method will allow for a small amount of air to escape while helping to avoid contamination thanks to the cotton filter you've created. Keep in mind that this will only be the needle and not the syringe itself. I personally prefer twisting the lid, but both methods work fine.

Pressure Cooker

With adequate oxygen allowance given to the jars, we can load them into the pressure cooker. It may help to cover your jars with aluminum foil or paper, as this will help to reduce the likelihood of moisture getting near the lid of the jar. There is no steadfast evidence that this actually works better, but it is better to be safe than sorry when it comes to potential contamination. If you prefer not to heat up aluminum foil, as some claim it is prone to being toxic, then paper will work fine.

Set your pressure cooker to 15 PSI and for a 15

minute cook time. It is important not to cook longer than this, as honey can sometimes begin to turn a brown, caramel color and make the process of determining if your bottle is contaminated much harder later on. Once the pressure cooking has completed, leave the jars in the pressure cooker for your second stint of waiting. After 24 hours, you will be ready to open up the pressure cooker and remove the jars.

If you used the method of allowing oxygen in by loosening the jar lid, just retighten it completely. At this point, you will need to make sure to remove any air pressure from within the jar, otherwise we'll have problems when introducing the spores into the liquid culture. To achieve this, make sure to sterilize the needle with the alcohol lamp, and then simply stick the syringe filled with alcohol-soaked cotton through the silicone that seals the hole on top of the lid. This cotton-alcohol filter helps to keep out potential contaminants. This will balance out the pressure within the jar to be identical to room pressure. If you used the method of sticking a needle into the top of the silicone to begin with, you can skip this step. In the event that your silicone area is not resealing itself, you may want to dab a bit extra on top at this point.

If you've taken precautions to keep your work area clean, yourself sterilized, and the jar itself sterilized, you should now have sterilized honey water that is almost ready for the mushroom spores.

Injecting the Spores

Finally, we're ready to add spores into our jars of honey water to create the liquid culture we need to grow our mushrooms. Start by first sterilizing the spore syringe's needle with the alcohol lamp. It is ideal that the needle becomes red hot during sterilization. Allow it to cool a bit before plunging it into the silicon opening. Once you place the needle into the silicon hole, there's no need to push it far enough down for it to make contact with the water. The spores can simply drop into the sterilized honey water solution. Allow them to settle a bit, and you will be able to stir the entire jar later simply by swishing it around and letting the marbles do their job.

The amount of spores you will want to inject can be as little as ¼ml, but it is advised to use 1ml or more for better results, especially as a beginner. Once again, when removing the needle, if the silicone does not close on its own, it is advised to have some additional silicone ready to go ahead and seal up the small hole to avoid contamination. You should be able to tell prior to this step if the silicone is not sealing properly, though.

What we have now is honey water and spores that are living in it. This is your liquid culture, and it will be the basis for your grow. Store this jar in a dark place, or at least out of reach of direct sunlight. Room temperature is fine for storage, something around 70-80 degrees Fahrenheit (20-26 Celsius) is ideal.

Here Comes the Mycelium

It is time to start waiting again! It can take a

week or two, but you should start seeing some white, or off-white, bits floating around in your jars. This is usually good news! That means that the mycelium has begun to propagate, and you're on your way to a potential yield. However, if the result ends up being a powdery mess or the color is off, there's a high chance that your jar has become contaminated, and it's time to dump it out and start off. For this reason, most cultivators really should be working with at least a few jars every time they create a liquid culture, as the smallest thing can cause contamination.

Throughout the growth of the mycelium, you will want to gently shake the jar so the marbles help to break up the mycelium into small enough pieces that it will easily be able to pass through the syringe needle. You will repeat this before using your liquid culture as well.

Once the substance is white and fluffy, you may very well be ready to start growing. To be safe, you can use a brown rice flour cake jar, as we'll discuss how to create shortly, to test out the liquid culture.

After using the method explained for the brown rice flour cake, give it a week or so to see if the growth of the mycelium remains white. It is highly advised not to skip this testing step, but you can always move forward without it and start making an entire batch of brown rice flour cakes with the liquid culture. You will waste some supplies and a lot of time and effort, but it won't be the end of the world.

When your liquid culture is fluffy and the mycelium is ready, you should store it in a cool place. Ideally, this would be in your fridge. It can last a few weeks outside of a fridge, but it will last several months if kept cool and away from light. Sometimes it can last as long as a year or more. Once you have this ready, you can store it until you're ready to create your brown rice flour cakes for growing.

Once you've determined that your liquid culture hasn't been contaminated, you are finally ready to put your liquid culture into a syringe and start the next step of the process. It is important, as always, that you maintain a sterile work environment. Many growers will use gloves during this stage to help aid in sterilization, but it isn't a requirement if you follow all the other steps correctly.

Remove your liquid culture from the fridge, and start by sterilizing the silicone area on the top of the jar lid. This is best done with alcohol, generally 70% alcohol is ideal in comparison to others. Once more, rotate the jar so the mycelium breaks up with the marbles that you've placed inside.

To ensure that the syringe is sterilized, especially if it is one that has been used before, make sure to clean out the syringe with rubbing alcohol. This is easily done by sucking in alcohol and sending it back into the bottle of alcohol. While you're doing this, have water boiling and at the ready. Suck this water into the syringe and discard the liquid. Repeat this a few times as well. This will remove the trace amounts of alcohol, which is important because alcohol could cause your liquid culture to die.

With the now empty syringe sterilized, we'll want to use the alcohol lamp again to sterilize the needle itself. Keep in mind that while you want the needle to become red, you do not need this to happen for an extended amount of time. Too much heat for too long of a duration can cause the syringe plastic to overheat and warp or melt completely. Again, avoid melting or damaging the plastic part of the syringe.

With the needle and syringe sterilized, we can now place the needle through the silicone on the liquid culture jar. Tilting the jar, make sure the hole in the needle is submerged in the honey water, and use the syringe to draw out the liquid culture. The small amount of 1ml (1cc) of this liquid culture is enough for one of your brown rice flour cake jars. Draw in this amount for each jar you plan to use. For 10 jars, 10ml (10cc) will suffice. A little extra definitely won't hurt, though.

Again, when dealing with the silicone, if it doesn't close itself back up automatically, you may want to dab a bit of extra silicone on your liquid culture jar to ensure that outside air isn't getting into your liquid culture. Because you'll likely have enough liquid culture to grow more than a single batch of mushrooms, it is important that you keep it sealed and in the fridge after this.

At this point, you should already be prepared to carry out the next steps for the least chances of contamination, so it is highly advised that you spend time understanding the rest of the process before creating your liquid culture.

Growing With Brown Rice Flour Cakes

We are assuming that you already have a liquid culture prepared and at the ready for your botanical journey growing magic mushrooms, and now it's time to begin the process of further growing the mycelium that eventually becomes mushrooms.

You will need quite a few things to use the brown rice flour cake method of growing your fungus.

1. **Mason jars** are the ideal vessel. The ½ pint mason jars with the wider openings that are popular in America are preferred, though other jars will work as well. In a punch, aluminum foil can be used to cover jars, but this is not recommended. With the internet, you should have no problem finding the perfect jars. It will be worth the extra costs, and you'll be able to use them over and over again.

2. **Vermiculite** is needed for growing. Vermiculite is a substance (aluminum-iron magnesium silicates) that is used to help with water-retention. This allows for the best possible environment for mushrooms to grow. It is wise to have both fine and coarse vermiculite on hand.

3. **Brown rice flour.** This will be part of the structure of our "cakes."

4. **Pressure cooker**. Larger ones are better because you can get more done, but most standard pressure cookers will work fine as long as they have easy-to-adjust settings.

5. **14-16 gauge syringes**

6. **Measuring tools**, such as cups and spoons are needed as well.

7. **Large mixing bowl**

8. **Ethyl-alcohol lamps can be used to easily sterilize needles.**

9. **Isopropyl (rubbing) alcohol. ideally 70%,** which is also used for sterilization purposes. It is highly recommended to stick to 70% alcohol.

You will need some additional supplies for creating the fruiting chamber where you will eventually grow your actual mushrooms, so be sure to check that section before you rush off to buy the required materials.

Jar Lids

If you are using mason jars, and you really should, then take the flat part of the lid and create four holes close to the very edges. These holes will need to be large enough to fit in your 14-16 gauge needles, so make sure they're sized appropriately. This can be done with a drill, but it is probably easier to use a nail and hammer, and it's a lot less messy. Make sure to have something underneath to ensure you're not putting a hole in your counter. You could use only two holes, but four holes allows for much quicker colonization in general. Because you'll likely be using a B+ strain, which grows slower anyway, using the four-hole method is may be more ideal. If your lid is a two-part lid, make sure you're not creating holes so close to the edge that the holes are covered when you put the lid together.

Pressure Cooker

You will fill the pressure cooker with roughly 5cm of water. Before placing your jars into this water, you will want to make sure that there's a way to keep the glass jars above the water. This can be achieved in many ways. Many growers use unused jar lids, as this is a simple solution and pretty much the perfect height. In place of jar lids, which could rust and become unusable later, you can use other materials. Often, you can place a towel in the bottom of the pressure cooker, and on top of this a glass serving plate or other large glass item with a base large enough for the amount of jars you plan to work with. However you do it, the glass jars need to be just out of the water when placed inside the pressure cooker.

Spawning Substrate

Your spawning substrate will be the home of your fungus. This is going to provide almost everything you need to allow your mushrooms to grow, so it is important that you follow these directions. This substrate will be comprised of:

- 2 parts vermiculite (roughly 3/4th cup per jar)

- 1 part water (roughly 1/3rd cup per jar)

- 1 part brown rice flour (roughly 1/3rd cup per jar)

Start out by mixing the vermiculite and the water in your large mixing bowl. For the brown rice flour method, it is best to use a coarse vermiculite. Add in water a little bit at a time until the vermiculite cannot absorb any more of the fluid but there also isn't a large amount of excess. Using a spoon or spatula, push the vermiculite down, and if there is a lot of fluid coming out, you've used too much water. You should only see a small amount of water leakage at most. If you must, start over on this stage. Pools of water will lead to mold problems.

Add in the brown rice flour to your vermiculite and water, and as you mix, it should begin making a fluffy consistency. It is absolute necessary to make sure there isn't excess water and that the water is added to the vermiculite before adding in the brown rice flour. If you do these in the wrong order or use too much water, you'll start to get a sticky paste, and that isn't what you want.

Once you have confirmed it to be fluffy and not dramatically sticky, you are ready to spoon this substrate into your jars. Loosely fill the jars leaving only 1cm of room for air at the top. There is no need to pack down this substrate, and leaving it loose will allow space for the mycelium to grow into the gaps in the loose substrate. Clean off the rim of the jar, and add in a layer of vermiculite on top. It is wise to use the finer vermiculite on top, as this acts as a filter against possible contaminants, especially mold. Add on the lid and close it tightly. You will need to repeat this process for each jar, loading them into the pressure cooker as you go. Cover the top of the jars with thick paper or aluminum foil to help keep moisture away from the lids as the pressure cooker begins.

Sterilization of the Jars

Now that you've made and added the substrate to each jar, it is time to sterilize everything. Making sure to check that you have water in the bottom of the pressure cooker and that your jars are above the water line first is a good idea. You will want to dial in your pressure cooker to 15 PSI. Allow this to heat on high heat, and once the pressure is reached, you'll need to lower the heat in order to maintain that pressure of 15 PSI. Try to keep this pressure the entire time, and allow the pressure cooker to continue for an hour. It is sometimes wise to test out your pressure cooker first with just water in it to learn how to best maintain this pressure. The goal is for the pressure cooker to not release any excess steam during the entire process. If steam releases, you'll likely need to start over, so a run with just water to figure out how to get the perfect pressure maintained is an easy way to avoid extra steps and frustration.

After an hour is up, your jars are now sterilized. It is best to leave them in the pressure cooker for 24 hours after this, but overnight is long enough. At this time, you should clean your work area up, and then go take a break from your experiment.

Adding the Liquid Culture

The next day, you should be able to continue

your journey. Take the pressure cooker into the room where you plan to add in your liquid culture, in our case a pre-prepared spore syringe. This room should have almost no air flow. Make sure to close the windows. Turn off any fans. Do not run the air conditioner. The less air flow, the better, as air flow can bring contaminants into your grow that can ruin all of your hard work. If you have or want to make a glove box, that is ideal, but it isn't necessarily a requirement.

Before injecting our spores into anything, we need to use the alcohol lamp's flame to sterilize our needle for our syringe filled with the liquid culture. It is ideal to do this before each jar, and you should also be sure to be wiping off the needle with alcohol to remove any wet vermiculite between each insertion.

We will add the liquid culture in our spore syringe to each of our jars through the holes we previously made in the lids. Keep in mind we really only need 1ml (1cc) of liquid culture, so 0.25ml-0.5ml for each of the four holes is fine. This doesn't have to be perfectly measured if you've allowed for extra liquid culture in your syringe.

When injecting the liquid culture into the jar, it is ideal to hold the jar at a tilt and allow the needle to be at an angle so the tip of the needle rests along the edges of the jar. You want it past the dry layer of vermiculite, as this should remain dry to avoid potential mold growth. You should be able to watch the liquid culture run down the side of the jar as you do this. While you could inject it straight down into the substrate, this method is ideal because it means that growth will start near the edges, allowing you to more accurately tell if there are any issues with contamination much quicker. You can really inject as much liquid culture as you want so far as there isn't a pool of liquid in the jar, so if you're ever unsure, just use a little bit more. That said, 0.25ml per hole for the four-hole method is plenty.

Repeat this for each jar.

The Colony Grows

Once you've finished adding in your liquid culture, you'll want to leave all of these jars on a shelf and keep them at room temperature. The ideal temperature is going to be below 80 degrees Fahrenheit (or 27 degrees Celsius). Higher temperatures may still allow your mycelium to colonize, but there's more likelihood of contamination.

The holes used to add in the liquid culture should not be plugged or have anything blocking them whatsoever. The mycelium will require that there is fresh air to grow. You can use paper on top of them to allow for breathing room while also helping to keep out the potential of germs or other contaminants, but this can be removed around the time your jar is half-filled with the white mycelium that tells you your colony has been growing. After two to three weeks, your jar should be almost entirely white with the mycelium colonization. The contents will almost become solid, other than the layer of vermiculite on the top. Hence the term "brown rice flour cake."

During the period of colonization, you will have time to prepare your **fruiting chamber**, which we'll cover shortly.

Rest In Peace? (Contamination)

Not every jar you add liquid culture to is guaranteed to be a winner. Unfortunately, you may have a few causalities along the way. This is one of the reasons that doing more jars at once is such a good idea. Obvious signs are discolored results. Your mycelium should be white, so if you are seeing green or brown growths, it's safe to assume there is contamination and the jar is a lost cause. Clean and then sterilize the jar, and make sure to dump it out far away from the others. If you see white that appears to be liquidy, it is most likely bacteria, and it's time to dump your jar. If you have multiple jars, you most likely have one that is doing well, so you should be able to tell the difference, but it is a bit harder than molds that are green.

In some instances, it is possible that a contaminant may start to grow inside the jar well after the mycelium is growing and thriving. Should you see this happen, you may want to give it a bit longer, as the mycelium can actually grow over molds and other contaminates. In this case, they will break down the contaminant, and your jar may well be useful still.

Fruiting Chamber

The "fruiting chamber" is essentially a makeshift terrarium. You can put this together while you are waiting on the colonization stage of your mycelium to finish, as it takes quite a while. After your first grow, you will be able to reuse your fruiting chamber.

The design for this particular piece of the equation can vary greatly, but the truth of the matter is that keeping it simple will often yield the best results anyway, so there's not really any sense in going overboard on this particular equipment. In fact, trying to automate any part of the process can prove to be futile because being hands-on and constantly evaluating your grow is going to help you combat any problems.

The most popular fruiting chamber design is known as a "shotgun fruiting chamber." The shotgun fruiting chamber is named as such because of the holes that will be drilled into it. To build and use this fruiting chamber, you will need the following supplies:

1. **Perlite**, and a lot of it. Perlite is a volcanic glass that helps allow air to pass through soil and other growing mediums.

2. **A clear tote or other transparent plastic container**, ideally rectangular and fairly large, will be used to grow your mushrooms in.

Transparency allows for indirect light, so it is crucial that you're not using anything that is opaque. The clearer, the better.

3. **You'll also want a drill that allows for hammer action to be completely disabled**, as hammer action can cause cracking when drilling into your plastic tub. (In a pinch, placing a block of wood on the other side of where you're drilling can minimize the likelihood of cracking the plastic). You will also want 6mm and 1-2mm bits.

4. **A spray bottle is needed.**

5. And lastly, you need some type of **lighting setup that will work with 6,000-7,000 kelvin fluorescent bulbs**, and a way to hook this up to be close to your grow setup. These bulbs should be similar to natural sunlight, and many brands may even advertise as such. Most 15w bulbs are fine.

On all six sides of your rectangular plastic tub,

you will drill 6mm (1/4th inch) holes roughly 5cm (2 inches) apart. It is important that you holes are no larger than this, as we need the fruiting chamber to maintain humidity in order for our mushrooms to grow. These holes should be drilled throughout the entire box, so along the height and the width of the box, top, lid, all sides, which is why it is called "shotgun fruiting chamber," as there will be quite a few holes.

Some growers will drill much smaller holes along the bottom of the tub, usually 1-2mm in size and use the normal 6mm size holes throughout the rest of the box. This design helps to keep the dampened perlite we'll add to the box from falling through. It will also allow you to clean the perlite after a successful flush or two has grown, keeping down future costs. If you attempt to rinse out the perlite and only use the larger holes, you'll find that you're just washing it down the drain. Realistically, perlite is cheap in many areas, so it is better to just replace it entirely, but if it isn't cheap or your budget is constrained, washing and reusing may be ideal for your usage.

You will also drill these larger 6mm holes into the lid of your container. The idea is that moisture can be maintained but there is also going to be airflow, which is crucial for growing healthy mushrooms. If your lid is not transparent, you will need to add your light source on the sides instead of the top, but a clear lid and a light source for the top is typically going to be easier. We'll cover this more shortly.

Returning to your tub, now with the holes drilled out throughout the entire box, you will want to add in 7-12cm of clean perlite. To be safe, you may want to wash the perlite in a colander first. This perlite should be wetted and added only when your brown rice flour cakes have been prepared and are ready to be put into your fruiting chamber. When doing this, you can start with a thin layer of dry perlite, 1-2cm only, and then add on the wetted down perlite afterwards. The moisture from the perlite on top will wet the rest of the perlite without drowning your fruiting chamber in too much liquid, which wouldn't be a big deal except there are holes in every part of your box.

Make sure to flatten out the layer of perlite in order to create ideal places for your cakes to set. This perlite will simply hold moisture within the box, and it is not going to actually do anything else, so the amount isn't totally critical and are just estimates. It really just depends on the size of the fruiting chamber. You should simply decide on the size of your box by determining how many cakes you intend to grow. To help make the decision easier, consider that a flush on a normal-sized cake, based on the directions here, will yield 2-7 grams. You'll likely want to do at least 5, but 10 or more is ideal.

Once you have completed this, it is time to place your fruiting chamber on an elevated surface that won't block the holes on the bottom of the box. This can be easily achieved with stones or bricks, as long as you can keep it secure on the table. Having an elevated location helps to reduce the likelihood of contaminants, as these tend to land near the ground, and it makes it easier to place a second, larger tub beneath to ensure there's no water drainage that can damage your table or home in any way. Again, it is important that the holes on the bottom of the box are exposed. These will allow for the necessary airflow required.

Once you have your box prepared, you will also need to prepare a light source. For most first timers, the ideal situation is going to be a light source that is above or to the side of their fruiting chamber. This can be as simple as a single bulb, and as stated, it should be within the 5,000-6,000 Kelvin range that help imitate natural sunlight that a mushroom would normally receive. Most 15W bulbs are appropriate. If possible, lighting from an angle is ideal, so putting together some type of rig to hold in the bulb slightly to the side is going to work well. The bulb should not touch the fruiting chamber whatsoever.

The actual setup isn't that important as long as there is light. When lighting your fruiting chamber, most growers give their grows 12 hours of "sunlight" and 12 hours of darkness per day. It is difficult to give your grow too much light, but keep in mind that the light should not be close enough to anything to scorch your mushrooms.

That's really all there is to the fruiting chamber and lighting setup. You can experiment with different setups, and plenty of people have, but this simple solution is a great way to ensure that you are giving your experiment the perfect living quarters.

Remember, do not add in your perlite until you're actually at the stage where you're getting ready to plant your brown rice flour cakes. This will only give it more time to become contaminated before you've even begun the final stages of growing. Everything else, however, are things you can handle while you're in the waiting stages of mycelium colonization.

Consolidation

When colonization is nearly complete, you will see that the jar is almost entirely white on the inside. This is a great sign. Make sure to look at the bottom of the jar to ensure it is truly complete, as this should also be white. At this stage, you should date each of the jars (I use a piece of blue painter tape, but anything is fine really). Give it one more week after this stage for good measure.

It is of no concern should the mycelium begin pulling the substrate away from the side of the jar, but should there be some small growths of mushrooms going on, that means you're already ready to get going and it's time to move onto the next step. You shouldn't see this happening too early in the process, though. This is pretty rare if you've taken care to keep the small holes on your lid a moderate size.

Once we've given it that last week, we can move forward by pulling out the mycelium, and then removing the dry vermiculite layers. These "cakes" should pull out of the jar without any issues and be rather dry. We will need to add moisture to allow for further growth. To do this, we simply plunge them into water for a 24-hour period. This can be done in a bucket or large bowl. You will want to put something over the cakes to ensure they remain in the water. A simple dinner plate usually does the trick, just make sure it's sterilized first.

Removing them from the water the next day, you will wash them in running water in the sink. They should still be very solid, so this shouldn't be causing any serious damage to the cakes.

It is now time to give these a new layer of vermiculite, which can be done by pouring it from a cup as you rotate them or filling a plate or another container with some dry vermiculite and rolling the cakes in the vermiculite. This should be a thin layer only. Fine vermiculite is ideal here for better coverage, but whatever vermiculite you have is going to work, as both will help to keep the cake moist.

Fruit(ing) Of Your Labor

As previously mentioned, you will need to have your fruiting chamber prepared already at this stage, and you should have handled while waiting for the mycelium to colonize the jars.

Taking these moist cakes that have been rolled in vermiculite, you'll place them within the moist perlite found in the fruiting chamber. For most strains of cubensis, you want to keep the temperature between 70-80 degrees Fahrenheit (23-27 degrees Celsius), though slightly lower temperatures are fine as long as you're patient and don't mind waiting longer for growth to occur.

Note that in colder climates, it is important to understand you cannot simply heat your fruiting chamber. You will need to find a way to heat the room itself to the ideal temperature range, otherwise you will very likely see damage and the potential for molds. You do not want direct light, and you do not want direct heat. These are harbingers of failure.

It is up to you, but some prefer to keep the cakes and perlite somewhat separated using plastic screen or aluminum foil, but honestly, putting it directly on the perlite is fine as well.

When adding in your cakes, you also have an option to stack some of these cakes together, which may benefit for a larger mushroom once grown, as the mycelium will gradually bind together. This is not necessary, and it is sometimes argued that smaller mushrooms will be more potent, but they are more of a hassle to harvest and prepare for consumption.

	You will need light to allow for growth of mushrooms. While we don't want direct light or heat, we want to have lighting setup for at least half a day, allowing for darkness the other half of the day. Once more, lighting at an angle is ideal, but directly above the fruiting chamber will work as well. As discussed, this lightning should mimic actual sunlight, so a lighting setup that will work with 6,000-7,000 kelvin fluorescent bulbs should do the trick. A very small amount of direct light, as in just 2-3 minutes per day, may help fruiting, but it shouldn't be necessary. Personally, I avoid this method, but many expert growers claim it helps.

	You also need to control the humidity. Mushrooms like high amounts of humidity. The goal is for at least 95% humidity within the fruiting chamber itself, which means that you want to use the water bottle to spray down your mycelium cakes around four times per day. If dryness is an issue where you live, or within the room you're using, then getting a humidifier is an easy solution to keep everything a little damp. The humidity within your room should be at least 30% for desired results.

During the first few days, do not mist the cakes themselves, but mist around them into the perlite within the fruiting chamber. Doing this ensures the vermiculite remains on the cakes, and from there, you may also want to lightly fan the cakes to allow for CO_2 displacement and allow fresh air to reach your grow. (I use a clean sheets of paper to make a fan. It's probably not necessary, but it seems more sterile to use something like this.) The end result of making sure humidification is perfect is that you have small amounts of water, like beads, but nothing close to pools or puddles. Remember, standing water is going to bring contamination and mold. If you're worried about water, you can keep a second plastic tub under the fruiting chamber to help keep water from getting all over the place.

Mushrooms Are Blooming

This takes a lot of work and effort, but finally you should start to see something coming up after a week or two of having your cakes in the fruiting chamber. You should see a mushroom partially veil. This "partial veil" is the membrane that appears under the caps. When these finally come through, you can harvest the mushroom. A gentle twist near the base is ideal.

Mushrooms will usually come in "flushes," so you won't be harvesting everything at one, but those that appear near the top are likely ready to go once they have a noticeable cap. This means that you will have the cap, but you will not see too much of the slotted-looking membrane from beneath the cap. This cap does not need to be huge, and taking the smaller ones as you take larger ones is likely going to be wise. If the veil has not broken on smaller ones, you may want to leave them to allow a bit more growth.

After a harvest is complete, you will likely be able to get another "flush" from your cakes. To make this possible, allow the cake to dry out for 24 hours. You can remove it from the fruiting chamber if other cakes are not quite finished yet. This helps to avoid potential for mold growth. After it has dried, you will again want to submerge the cake in water for 12-24 hours (less time is fine for the second flush) before placing it back into the fruiting chamber to begin the next flush. There's no need to rinse or add the vermiculite again.

The first flush will almost always bear the best results, and while some claim to get multiple flushes per cake, getting two decent harvests is a huge success. You can try to get as many flushes as possible, but at some point, the cake is essentially only good for composting.

Fruiting Gone Bad

In some cases, there are tell-tale signs during the fruiting process that something isn't quite right. An immediate concern will be fuzziness on the stems, typically white in color. This is a sure sign that there needs to be better lighting or fresh air for their growth.

Smaller mushrooms from a first flush may suggest that the consolidation period wasn't as long as it should have been, or there isn't as much humidity or light as the mushrooms truly need to thrive. When a mushroom stops growing for this reason, there's a good chance it's time to just pick them. If all the mushrooms on a cake are tiny but ready to harvest, that means you've used the cake too many times or there's far too much moisture. In some instances, they may be contaminated.

A blue streak may be present on your mushroom, which is usually near the edge of the mushroom's cap, and that means it's not likely to grow any more.

If you don't see anything to suggest this otherwise (moisture levels, lack of light), then you need to look for mold. This bluish streak is a good sign that you want to remove the cake completely, as mold can spread from one cake to another.

If none of this has happened, congratulations! You've got mushrooms!

Harvested... Now What?

At this point, you have already went through the process of creating a liquid culture, colonization of the mycelium, and growing mushrooms. You've successfully grown magic mushrooms, and now it's time to prepare them for consumption and storage.

Before doing anything, it is probably wise to go ahead and clean these mushrooms off. This can be done simply with water. It is up to you if you're worried about the vermiculite and perlite, but if you are, go ahead and cut off the bottom of the stems a little bit as well.

There are a two very common ways to dry your mushrooms. The easiest method is probably to put them on a drying rack in a dry room (so not in the room where you've been running a humidifier). Using a fan as your air source, they should dry out within a few days. In particularly dry climates, they may be ready in as little as one day. When they are dry enough, they'll be almost like a cracker. If they bend when you try to snap them, they aren't ready yet. With this method, the potency of your mushrooms will last a few months, so they still need to be consumed in a fairly short amount of time.

For those growing larger amounts or wanting to store their mushrooms for much longer, you will still want to dry them on a drying rack, but you can take the process a step further by using **desiccant** to further dry your mushrooms. (It is not wise to just use desiccant, as this is a waste of resources and takes far too long on its own.) After they have dried, you can use a desiccant like calcium chloride and place it into an airtight container with two simple mesh surfaces made of plastic. These mesh surfaces should be one above the other. With these, you will have three chambers in your container. The bottom is for water runoff, and you won't have to place anything there. The middle is where the calcium chloride (or other desiccant) is placed, and the top is where you mushrooms will go. Make sure that the top two layers do not touch whatsoever. This process will dry out your mushrooms further, and it will allow them a much longer shelf life. Note that they can simply be stored this way if you wish to help reduce the likelihood of new moisture being introduced to your mushrooms.

Consumption and Final Thoughts

Finally, we're ready to partake. It has been a long journey to this point, so you should revel in the opportunity to enjoy yourself to the fullest.

These magical mushrooms can be consumed fresh directly after harvest and a wash, but considering that mushrooms are almost entirely comprised of water, you will have to eat quite a bit of these slimy morsels. The amount you need depends largely on potency and your own body chemistry, so I can't offer a realistic guideline for the amount to consume.

If you're familiar with dried mushrooms, consider that 20 grams of fresh mushrooms may be as little as 2 grams of dried mushrooms. For most people 1-2 grams of dry mushrooms as a starter dosage is plenty. If taste and amount is a concern, then drying method is probably going to be better for you. Since you likely won't eat an entire harvest anyway, and these types of mushrooms can rot quickly, you will almost inevitably be drying at least some of every flush that you harvest.

There are sometimes issues with nausea that comes with eating these magic mushrooms. Some people don't experience this at all, though. Marijuana is a common solution to help alleviate the nausea, though many claim that it will also intensify the other effects. Other solutions include eating oranges or other citrus prior to ingestion, and even better, blending the mushrooms into citrus-based juice and drinking it this way. There are many other ways to prepare shrooms, including teas, and likewise, many cultures have their own rituals, such as chewing on mushrooms for long periods of time. You can experiment.

While there is a lot to say about the consumption of magic mushrooms, there are very few people with the literary, spiritual, or scientific knowledge that can truly explain what it is like to take a trip with your new friend psilocybe cubensis. Reports and anecdotal information is an excellent resource to alleviate the pre-consumption jitters, but body chemistry, environment, potency, comfort level, intelligence, and a huge number of other factors can dramatically change not only how you feel after consumption, but also how you process the events that occurred during consumption and what they will mean to you long after the experiment has ended. It can be a night of just having fun just as easily as it can lead to a spiritual awakening (even if that spiritual awakening isn't going to be taken to the extreme afterwards).

For myself, it suffices to say that there is a reason so many people have found spiritual properties during their time spent with magic mushrooms. Even those without a spiritual bent should be able to appreciate this and consider the fact that this isn't necessarily just a way to "party." Many might go so far as to claim that it isn't even remotely ideal for a "party" situation. Naturally, this is something for you to determine on your own. Chances are, if you are feeling comfortable, you will have fun if you eat mushrooms during a music festival, but this experience can be vastly different than eating them in the comfort of your home with a close friend.

If you are completely new, then my advice is to just take it slow. There's no reason to race toward nirvana, especially as a grower that will have a new harvest before too long. There will always be more mushrooms and more chances to experiment. Best of luck in your endeavors.

www.ingramcontent.com/pod-product-compliance
Lightning Source LLC
Chambersburg PA
CBHW020709180526
45163CB00008B/2999